GRE
Formula Book

GRE
Formula Book

Saifuddin Kamran

To order additional copies of this book, contact:
Xlibris
AU TFN: 1 800 844 927 (Toll Free inside Australia)
AU Local: 0283 108 187 (+61 2 8310 8187 from outside Australia)
www.Xlibris.com.au
Orders@Xlibris.com.au
820740

CONTENTS

1. FRACTIONS

Addition of fractions

Addition of two fractions

1) When the denominators are the same, add the numerators

$$\frac{5}{11} + \frac{3}{11} = \frac{8}{11}$$

2) When the denominators can be made same by simple multiplication

$$\frac{5}{18} + \frac{1}{6}$$

$$\frac{5}{18} + \frac{3}{18} = \frac{8}{18} = \frac{4}{9}$$

3) When the numerator is 1 in both fractions

$$\frac{1}{4} + \frac{1}{13}$$

Add the denominators for the numerator of the result, $13 + 4 = 17$
Multiply the denominators for the denominator of the result, $13 \times 4 = 52$

$$\frac{17}{52}$$

4) When none of the above conditions fit, then cross multiply and add the results for the numerator and multiply both denominators for the resultant denominator.

$$\frac{5}{11} + \frac{3}{7}$$

$$= \frac{5}{11} \times \frac{3}{7}$$

$$= \frac{5(7) + 3(11)}{11 \times 7} = \frac{68}{77}$$

Subtraction of fractions:

Similar to the above procedure

Multiplication of fractions:

In case of multiplication, any number on the numerators can be divided by (cancel proportionately) any number on the denominators.

$$\frac{5}{8} \times \frac{4}{21} \times \frac{42}{55}$$

$$= \frac{\cancel{5}}{8} \times \frac{4}{\cancel{21}} \times \frac{\cancel{42}}{\cancel{55}}$$

$$= \frac{1}{1} \times \frac{1}{1} \times \frac{1}{11} = \frac{1}{11}$$

Practice:

$$3\frac{5}{7} \times 1\frac{1}{2} \times 2\frac{2}{13}$$

Division of fractions:

For division of fractions, flip the dividing fraction after switching the division sign with that of multiplication.

$$\frac{4}{9} \div \frac{8}{11}$$

$$\text{Becomes } \frac{4}{9} \times \frac{11}{8}$$

$$\frac{4}{9} \times \frac{11}{8} = \frac{11}{18}$$

Practice:

$$4\frac{2}{17} \div 1\frac{3}{17}$$

---

2. DECIMALS

Types of decimals:

Every rational number is either a terminating decimal or a recurring decimal. A fraction in lowest terms with a prime denominator other than 2 or 5 always produces a repeating (recurring) decimal.

Note that 22/7 is a rational number but π is not, as 22/7 is an approximation of the irrational number π.

1) *Terminating decimals:*

As the name suggests, in this case, the decimal value terminates after a certain number of digits. 15.58, 3.232, etc. are terminating decimals.

A fraction results in a terminating decimal only if the numerator is divided by any multiple of 2 and/or 5.

$$\frac{13}{50} = \frac{13}{2 \times 5^2} = 0.26$$

2) *Recurring decimals:*

Also expressed by a bar on the recurring numbers, these decimals do not terminate but are continuous infinitely.

$$\frac{1}{3} = 0.\overline{3} = 0.333333...$$

$$\frac{3}{7} = 0.\overline{428571} = 0.42857142857......$$

Percentages are fractions with a base value of 100. So 3/5, that is equal to 60/100, can also be stated as 60%.

CONVERTING FRACTIONS TO PERCENTAGES:

Multiply the fraction by 100.

For example $\frac{4}{5}$, when converted to percentages becomes

$$\frac{4}{5} \times 100 = 80\%$$

CONVERTING PERCENTAGES TO FRACTIONS

Divide by 100.

$$75\% \text{ becomes } \frac{75}{100} = \frac{3}{4}$$

CONVERTING PERCENTAGES TO DECIMALS

Divide by 100.

$$65\% \text{ becomes } \frac{65}{100} \text{ or } 0.65$$

FINDING CHANGE PERCENT

$$Change \% = \frac{Change}{Original} \times 100$$

FINDING PROFIT OR LOSS

Note that, unless mentioned otherwise, profit or loss is calculated with respect to cost price of a product.

$$Profit\% = \frac{Profit}{Cost\ Price} \times 100$$

$$Loss\% = \frac{Loss}{Cost\ Price} \times 100$$

FINDING SELLING PRICE

$$Selling\ Price = (100 \pm x)\% \times Cost\ Price$$

Where + is used in case of profit and − is used in case of loss.

4. PRIMES

Prime numbers are those numbers that have exactly two divisors. These two divisors are 1 and the number itself. A prime number cannot be written as the product of two factors, both greater than 1. In other words, primes are integers greater than one with no positive divisors besides one and itself. Prime numbers can never be negative. All other numbers, other than 1, are called Composite numbers.

Prime numbers: 2, 3, 5, 7, 11, 13, 17, 19, 23, 29, 31, 37, 41, 47, ...

Composite numbers: 4, 6, 8, 9, 10, 12, 14, 15, 16, 18, 20, 21, 22, 24, 25, 26, 27, 28, 30, ...

- 2 is the only even prime number. 2 is also the smallest prime number.
- After 2 and 5, all following prime numbers end only in 1, 3, 7, and 9.
- All prime numbers other than 2 and 3 can be written in the form $6m\pm1$, where m is a positive integer other than zero.
- All non-zero natural numbers can be factorized as products of two or more prime factors.

5. QUADRATIC EQUATIONS

SOLVING QUADRATIC EQUATIONS:

The standard form of a Quadratic equation is

$$ax^2 + bx + c = 0$$

To solve a quadratic equation, first multiply a by c.

Now, think of two numbers whose product is ac and whose sum is b. Let those two numbers be m and n.

When $a = 1$

Then the original quadratic equation will be factorised as follows:

$$ax^2 + bx + c = 0$$

$$=> (x + m)(x + n) = 0$$

When $a > 1$

Then the original quadratic equation will be factorised as follows:

$$ax^2 + bx + c = 0$$

$$=> (ax + m)(x + n/a) = 0$$

QUADRATIC FORMULA:

When
$$ax^2 + bx + c = 0$$

Then

$$x = \frac{-b \pm \sqrt{b^2 - 4ac}}{2a}$$

Po-Shen Loh's method of solving quadratic equations

$$x^2 - 4x - 12 = 0$$

The two roots will be $x = \frac{-b}{2a} \pm z$,

where z can be found by using the equation $\frac{b^2}{4a^2} - z^2 = c$

In this case $\frac{(-4)^2}{4(1)^2} - z^2 = -12$

$$\therefore z^2 = 4 + 12$$
$$z = 4$$

So the roots are

$$x = \frac{-(-4)}{2(1)} \pm 4$$

i.e. $x = 6 \ or -2$

6. NUMBER PROPERTIES

Even and Odd numbers

Integers can be Odd or Even.

An even integer is one that is divisible by 2.

Zero is an even integer.

Even and odd integers can be positive or negative.

Odd numbers: …,-5, -3, -1, 1, 3, 5,…

Even numbers: …, -4, -2, 0, 2, 4, …

Product of any non-zero number with an even number is an even number.

Product of any two odd numbers is an odd number.

Sum of two even numbers or two odd numbers will always give an even result.

Sum of one odd and one even number will always give an odd result.

Even numbers can be written in the form $n = 2k$ where k is an integer. It means even numbers are perfectly divided by 2.

Odd numbers can be written in the form $n = 2k + 1$. It means odd numbers leave a remainder of 1 when divided by 2.

7. PROPERTIES OF ZERO AND ONE

ZERO

1) Zero is an integer.
2) Zero is an even number.
3) Zero is neither positive nor negative.
4) Zero is not a prime number.
5) The difference of two equal numbers is zero.
6) Zero is called an Additive Identity because any real number added to zero does not change the number.
7) Zero multiplied by any number converts it to zero.
8) The product of two or more numbers can only be zero if at least one of them is equal to zero.
9) Any number to the power zero, is equal to 1.
10) Zero divided by any non-zero number is equal to zero.
11) The number zero, raised to any power, equals zero.
12) Dividing any number by zero is not possible. Therefore, zero can never be the denominator of a fraction.

ONE

1) 1 is the Multiplicative Identity because any number multiplied by 1 will retain its original value.
2) 1 is neither a prime number nor a composite number.
3) Any number to the power 1, is still the same number.
4) Any number to the power zero, is equal to 1.
5) The number one, raised to any power, equals one.
6) Any nonzero number divided by itself equals one.

8. DIVISIBILITY RULES

If a number is divisible by

2: The last digit is 0, 2, 4, 6, and 8

3: The sum of all digits is divisible by 3

4: The last two digits are 00 or are divisible by 4

5: The last digit is either 5 or 0

6: The number is divisible by BOTH 2 and 3

7: The difference between twice the last digit and the remaining number is divisible by 7

8: The last three digits are 000 or are divisible by 8

9: The sum of all digits is divisible by 9

10: The last digit is 0

11: The difference between the sums of alternate digits is 0 or a multiple of 11

12: The number is divisible by BOTH 3 and 4

9. LCM AND GCF

A factor or a divisor d, of a non-zero integer x, divides x into f integers, without leaving any remainder. Mathematically,

$$x = \mathbf{fd} \text{ (Where all x, f, and d are all integers.)}$$

All integers are factors of themselves.

GCF (Greatest Common Factor) or HCF (Highest Common Factor) of two or more non-zero integers is the largest integer that divides those numbers evenly. For example, the GCF of 12 and 32 is 4, and the GCF of 15, 12, and 18 is 3.

LCM (Least Common Multiple) of two or more non-zero integers is the smallest integer that can be divided evenly by any of those numbers. For example, the LCM of 12 and 32 is 96, and the LCM of 15, 12, and 18 is 180.

The product of two numbers is equal to the product of their LCM and GCF.

10. ROOTS AND EXPONENTS

When expressed as a^n, n stands for the number of times a is multiplied, where a is called the base and n the power or exponent. For example, 4^3 is equal to $4 \times 4 \times 4 = 64$

Some important facts about exponents:

$a^0 = 1$ – Any non-zero number to the power zero is equal to 1.

$a^1 = a$ – Any number to the power 1 is equal to the number itself.

$0^n = 0$ – When $n > 0$, means when n is positive. When $n < 0$, 0^n becomes undefined.

$1^n = 1$ – One multiplied by itself will always result in 1.

$a^{-1} = \dfrac{1}{a}$ and $a^{-n} = \dfrac{1}{a^n}$

$x^a \times x^b = x^{a+b}$ – When bases are same, the powers are added in case of multiplication.

$\dfrac{x^a}{x^b} = x^{a-b}$– When the bases are same, the powers are subtracted in case of division.

$x^a \times y^a = (xy)^a$ - When the powers are same, the bases multiply in case of multiplication.

$\dfrac{x^a}{y^a} = \left(\dfrac{x}{y}\right)^a$– When the powers are same, the bases divide in case of division.

$(x^a)^b = x^{ab}$- Successive powers get multiplied.

$$\sqrt[a]{x} = x^{\frac{1}{a}} \text{ and } \sqrt[b]{x^a} = x^{\frac{a}{b}}$$

$$\sqrt{x^2} = |x|$$

$x^a - y^a$ is divisible by $x + y$, if a is even and not divisible if a is odd

$x^a + y^a$ is divisible by $x + y$, if a is odd, and not divisible if a is even

11. INEQUALITIES AND ABSOLUTE VALUES

INEQUALITIES

When two mathematical expressions are equal, the relationship between them is called an equation. Whereas when the two terms are not equal then the relationship is called an inequality. Inequalities are solved in the same manner as the equalities with some exceptions.

Inequalities are expressed by symbols $\neq$ (is not equal to), $<$ (is less than), $\leq$ (is less than or equal to), $>$ (is greater than), or $\geq$ (is greater than or equal to).

e.g. $3x + 4 > 17$

$3x > 13$
$x > \dfrac{13}{3}$
$x > 4.33$

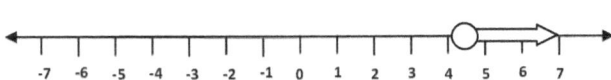

$7x - 10 \leq 2x + 11$
$5x \leq 21\text{w}$

$x \leq \dfrac{12}{5}$
$x \leq 4.2$

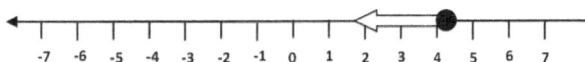

ABSOLUTE VALUE

The absolute value or Modulus of a certain number x, denoted by $|x|$, is its distance from zero. So whatever is in the modulus comes out as a positive value.

$|4| = 4$ or $|-5| = 5$

Remember that

$\sqrt{x^2} = |x|$

$|x| + |y| \geq |x + y|$

$|x| - |y| \leq |x - y|$

12. RATIO AND PROPORTION

The ratio of the number m to another number n can be represented as any of the following:

$$m \text{ to } n, \text{ or}$$
$$m: n, \text{ or}$$
$$\frac{m}{n}$$

Keep in mind that m: n is not the same as n: m. In fact it is the reciprocal of the other ratio.

A Proportion is a relationship equating two ratios.

So a:b :: c: d means

$$\frac{a}{b} = \frac{c}{d}$$

Solving Ratio and Proportion problems

Direct Proportion

16 boys spend an average of $320 in a week. At the same rate, how much will be spent by 21 boys?

In case of Direct Proportions, we use this path X

$$
\begin{array}{cc}
\textbf{Boys} & \textbf{Dollars} \\
16 & 320 \\
21 & x
\end{array}
$$

Following the path,

$$16x = 320 \times 21$$
$$\therefore x = \$420$$

Inverse Proportion

16 workmen can erect a wall in 10 days. In how many days can 8 such workmen erect a similar wall?

In case of Inverse Proportions, we use this path =

Workmen	Days
16 ———	10
8 ———	x

Following the path,

$$16 \times 10 = 8x$$
$$\therefore x = 20 \text{ days}$$

SOME INTERESTING RELATIONSHIPS:

If $\dfrac{a}{b} = \dfrac{c}{d}$, then

$$\frac{a+b}{b} = \frac{c+d}{b} \ (Componendo),$$

$$\frac{a-b}{b} = \frac{c-d}{b} \ (Dividendo),$$

$$and \ \frac{a+b}{a-b} = \frac{c+d}{c-d} \ (Componendo \ and \ Dividendo)$$

13. WORK AND RATE PROBLEMS

If Alexis can do a certain job in 20 days and David can do the same job in 30 days, in how many days will they be able to finish the job if they work together?

As Alexis takes 20 days to do the job, work done in one day by Alexis will be $\frac{1}{20}$

Similarly, work done in one day by David will be $\frac{1}{30}$

If they work together, the work done in one day will be equal to $\frac{1}{20} + \frac{1}{30} = \frac{1}{12}$

Therefore the work will take 12 days to finish.

METHOD:

Add the reciprocals and then flip the result to get the final answer.

Machine A can fill a certain container in 10 hours and Machine B can fill the same container in 15 hours. In how many hours will the container be filled if both machines are used simultaneously?

Adding the reciprocals:

$$\frac{1}{10} + \frac{1}{15} = \frac{1}{6}$$

Flipping the result:

$$\frac{1}{6} = 6 \text{ hours}$$

---

14. INTEREST PROBLEMS

Simple Interest:

When interest is applied on the original investment only, it is called Simple Interest. It can be calculated using

$$S.I. = PRT$$

Where P is the Principal Sum, R is the rate in per cent per annum, and T is the time in years. The final amount after T years will thus be equal to

$$A = P + S.I.$$

Compound Interest:

When interest is also applied on the interest earned in addition to the original investment, it is called Compound Interest.

In case of Compound Interest, the final amount is calculated using the formula

$$Final\ Amount = P(1 + R)^n$$

Where P is the Principal sum, R is the rate in per cent per annum, and n is the number of years.

In case of compounding quarterly

$$Final\ Amount = P\left(1 + \frac{R}{4}\right)^{4n}.$$

In case of compounding monthly

$$Final\ Amount = P\left(1 + \frac{R}{12}\right)^{12n}$$

15. AVERAGE

The average of n terms is given by

$$A = \frac{Sum\ of\ n\ terms}{n}$$

In case of a set of an odd number of equally distanced terms, the middle term is the average. So, in case of 7 equally spaced terms, the 4th one is equal to the average of the 7 terms.

$$6, 9, 12, 15, 18, 21, 24$$

$$\uparrow$$

Average

(Alternatively, the average of the first and the last term is also equal to the average of the series.)

$$A = \frac{6 + 24}{2} = 15$$

In case of a set of an even number of equally distanced terms, the average of the middle two is the average of all terms. For example, in case of 8 equally spaced terms, the average of 4th and 5th terms is equal to the average of the 8 terms.

$$6, 9, 12, 15, 18, 21, 24, 27$$

$$\uparrow$$

Average of 15 and 18 = 16.5

(Alternatively, the average of the first and the last term is also equal to the average of the series.)

$$A = \frac{6 + 27}{2} = 16.5$$

The average of a set of a numbers can never be smaller than the smallest term in the set and can never be greater than the greatest term in the set.

Addition of a term greater than the average of a set of numbers will raise the average of the set and addition of a term smaller than the average of a set of numbers will lower the average of the set.

Addition of a term equal to the average of a set of numbers will not change the average of the set.

16. MEAN, MODE, MEDIAN

Arithmetic Mean, or simply the mean, is the same as the average.

Mode of a set of numbers is the number that appears most frequently in the set.

Median of a set of numbers is the middle number when the numbers are arranged in order.

In a set of odd number of terms, $\dfrac{n+1}{2}$ term is the middle term.

In a set of even number of terms, the average of $\dfrac{n}{2}$ and $\dfrac{n+2}{2}$ terms is the middle term.

In an equally spaced set of numbers, the mean of the set is equal to the median of the set.

Range of a set of numbers is the difference between the largest and the smallest number in the set.

Standard Deviation of a set of numbers is given by

$$S.\,D.=\sqrt{\dfrac{\Sigma|x-\mu|^2}{N}}$$

Where $|x-\mu|$ is the difference between the mean and the value, and N is the number of terms.

A. Adding or subtracting a constant to each term in a set will not change the Standard Deviation of the set.
B. Multiplying or dividing by a constant to each term of a set will multiply or divide the Standard Deviation by that constant.

17. COUNTING PROBLEMS

Counting problems involve finding out total number of ways something can be done. The easiest form of a counting problem is one in which we just multiply the number of options to get the number of total possibilities.

How many suits can be developed from 3 shirts and 4 pants?

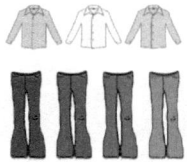

Here each shirt can go with any of the four pants.

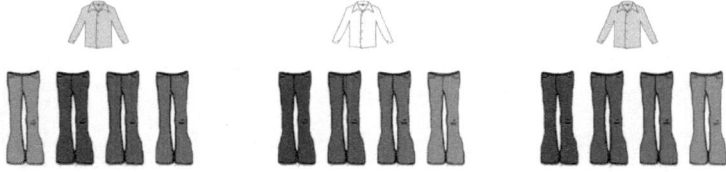

Hence there are 3 x 4 = **12 suits** possible

In how many ways can a worker enter and exit a building if there are 4 doors and she always enters and exits from a different door?

Entry options: 4

Exit options: 3 (One less)

Total options: 4 x 3 = **12**

How many possible combinations of a 4-digit lock can be can be made if all digits 0 to 9 are available to be used?

10 options for each place

Total combinations: 10 x 10 x 10 x 10 = 10,000

How many possible combinations of a 4-digit lock can be can be made if all digits have to be different?

First place: 10 options

Second place: 9 options and so on

Total combinations: 10 x 9 x 8 x 7 = **5040**

While standing in a line Darius notes that 15 persons were standing ahead of him and 16 behind him. How many persons were there in the line?

16	Darius	15

Total number of persons = 16 + 1 + 15 = **32**

In a line to buy tickets for the show, Sheila noted that she was the 7ᵗʰ person on the line. If there were a total of 24 persons in the line, how many were after Sheila?

	7	
	7	6
24-1-6	7	6

17	7	6

There were 17 persons standing after Sheila in the line.

George is 8th person in a line and John is 21st. How many persons are there between George and John?

21	8

The number of persons between George and John are $21 - 8 - 1 = \underline{\textbf{12}}$

(Note that when end-points are not to be included in counting we **subtract** 1 from the difference!)

In a charity drive, Harry sold tickets numbered 8093 to 8156. How many tickets did he sell?

Number of tickets sold = $8156 - 8093 + 1 = \underline{\textbf{64}}$

(Note that when end-points are included in counting we **add** 1 to the difference!)

How many multiples of 7 are there between 23 and 300, inclusive?

The first multiple that comes between the given numbers is 28 (4 x 7).

For the last one, divide 300 by 7. We get 42r6. So the last one is 42. Total number of multiples are $42 - 4 + 1 = \underline{\textbf{39}}$

18. PERMUTATIONS AND COMBINATIONS

Factorials:

The product of first n numbers is called n factorial, denoted by n!, provided that n is a positive whole number.

The last digit of all factorials greater than 4 is zero.

$0! = 1$

$1! = 1$

Permutations:

r items taken from n items and placed in a straight line. Order matters.

$$^nP_r = \frac{n!}{(n-r)!} \ (0 \leq r \leq n)$$

Combinations:

r items taken from n items and placed in a bag. Order does not matter.

$$^nC_r = \frac{n!}{r!(n-r)!} \ (0 \leq r \leq n)$$

$^0C_0 = 1$

$^nC_0 = 1$

$^nC_n = 1$

$^nC_r = {}^nC_{n-r}$

If $^nC_r = {}^nC_k,$ then $r = k,$ or $n - r = k$

$^nC_r = {}^{n-1}C_{r-1} + {}^{n-1}C_r$

If n is even, ^nC_r is greatest for $r = \dfrac{n}{2}$

Number of ways n items can be arranged in a straight line = $n!$

Number of ways n items can be arranged in a circular path = $(n-1)!$

Number of ways n items can be arranged in a circular ring that can be lifted = $\dfrac{(n-1)!}{2}$

19. PROBABILITY

$$Probability\ of\ an\ event = \frac{Number\ of\ desired\ outcomes}{Total\ number\ of\ possible\ outcomes}$$

(Condition: The outcomes are all equally likely to happen)

- Probability is always positive.
- Probability is always between 0 and 1.
- Zero probability means no chance of the event ever happening.
- Probability equal to 1 means no chance of the event not happening.
- P of A happening = 1 – (P of A not happening)
- The odds of an event is the ratio of its probability that it will happen to the probability that it will not happen.
- P(A or B) = P(A) + P(B) for Disjoint events
- P(A or B) = P(A) + P(B) – P(A and B) for Overlapping events
- P(A and B) = P(A) . P(B)

20. GEOMETRY – LINES AND ANGLES

A straight line has 180 degrees.

Supplementary angles add up to 180 degrees.

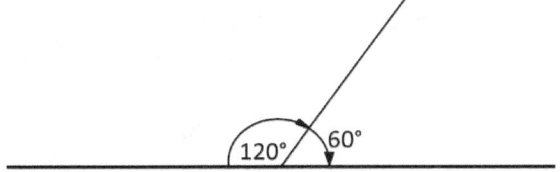

Complementary angles add up to 90 degrees.

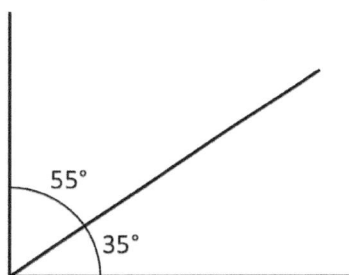

Two straight lines that make a right angle are perpendicular to each other.

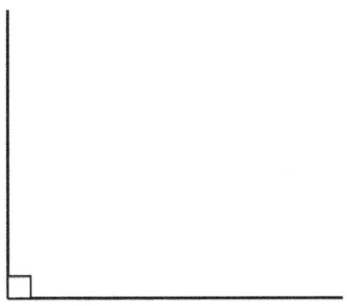

21. PARALLEL LINES:

Lines A and B are parallel.

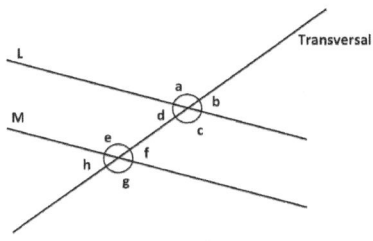

∠a=∠c or ∠b=∠d   (Vertical angles are equal)
∠b=∠f or ∠c=∠g   (Corresponding angles are equal)
∠a+∠b = ∠b+∠c = 180° (Adjacent angles are supplementary)

Parallel lines have the same slope.

TYPES OF ANGLES

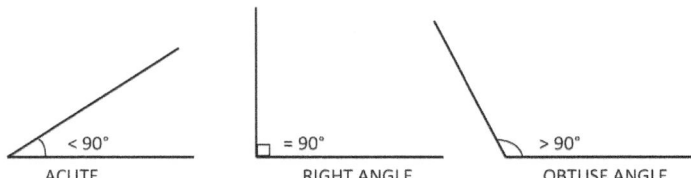

SPECIAL CASE:

If **A||B,** then the internal angle between the lines is the sum of the two internal angles.

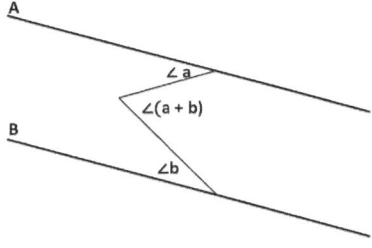

22. GEOMETRY – TRIANGLES

Sum of the lengths of two sides of a triangle is greater than the length of the third side.

$$\overline{A} + \overline{B} > \overline{C}$$

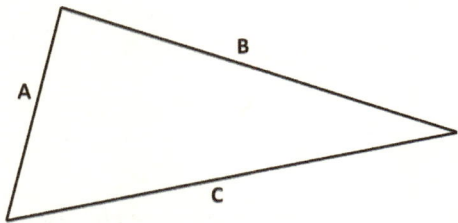

Sum of the angles of a triangle is equal to 180°.

$$\angle a + \angle b + \angle c = 180°$$

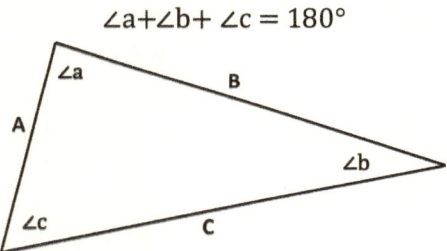

The exterior angle of a triangle is equal to the sum of the two opposite interior angles.

$$\angle d = \angle a + \angle b$$

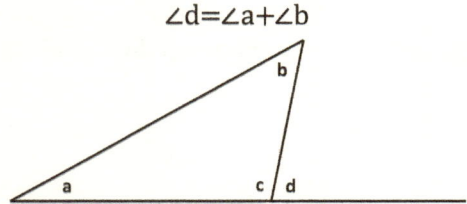

AREA OF A TRIANGLE

Area = Half of the multiple of two perpendicular lengths.

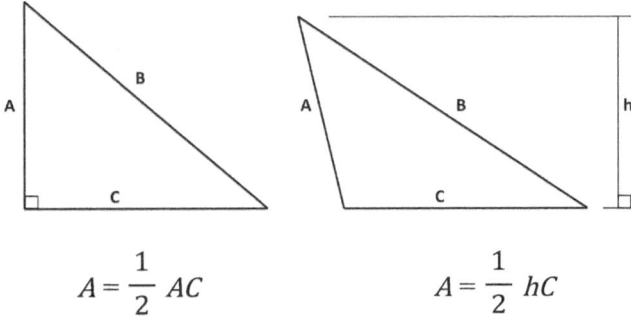

$$A = \frac{1}{2} AC$$

$$A = \frac{1}{2} hC$$

CLASSIFICATION OF TRIANGLES (Based on sides):

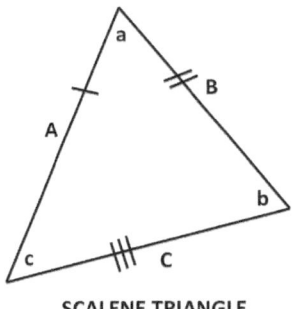

SCALENE TRIANGLE

SCALENE TRIANGLE
7. No two sides are equal.
8. No two angles are equal.
9. The angle opposite the longest side is the greatest angle.
10. It has no line of symmetry.
11. In a scalene obtuse triangle, the circumcenter will lie outside the triangle.

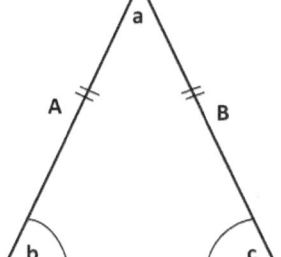

ISOSCELES TRIANGLE

ISOSCELES TRIANGLE
12. Two sides of an isosceles triangle are equal.
13. Two angles of an isosceles triangle are also equal.
14. A line of symmetry can be drawn that is perpendicular to the unequal side.

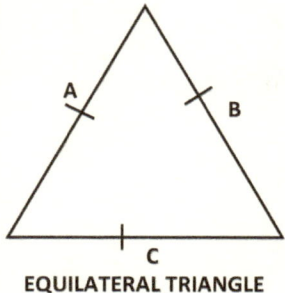

EQUILATERAL TRIANGLE

EQUILATERAL TRIANGLE

15. An equilateral triangle has all its sides equal.
16. All angles of an equilateral triangle are equal to 60°.
17. Altitudes drawn from all vertices bisect the triangle.
18. $Area = \dfrac{x^2\sqrt{2}}{4}$, where x is the length of the side.

CLASSIFICATION OF TRIANGLES (Based on angles):

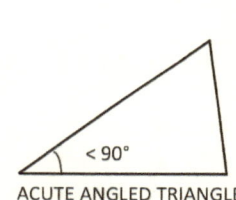

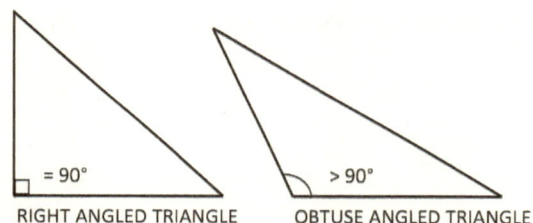

ACUTE ANGLED TRIANGLE　　RIGHT ANGLED TRIANGLE　　OBTUSE ANGLED TRIANGLE

PYTHAGORAS THEOREM

The sides of a right angled triangle follow a theorem called Pythagoras Theorem. According to this theorem, the square of the longest side (the hypotenuse) is equal to the sum of the squares of the other two sides (the base and the perpendicular).

$$C^2 = A^2 + B^2$$

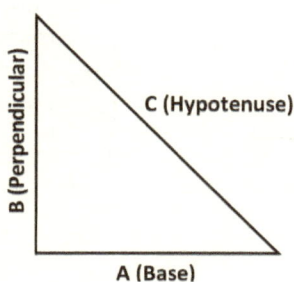

Some common right triangles:

Based on sides:

3 – 4 – 5 Triangle

5 – 12 – 13 Triangle

8 – 15 – 17 Triangle

7 – 24 – 25 Triangle

Based on angles:

A. **45° – 90° – 45° Triangle** The base = the perpendicular

The hypotenuse = $\sqrt{2}$ times the base Area of triangle = $\dfrac{1}{2}x^2$

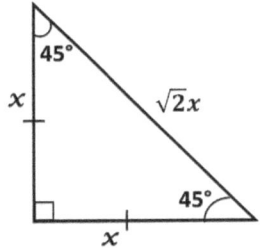

B. **30° – 60° – 90° Triangle**

The side opposite the 30° angle is the shortest.

The side opposite the 90° angle (The hypotenuse) is the longest and is twice in length to the shortest side.

The three sides are in the ratio $x: \sqrt{3}x: 2x$

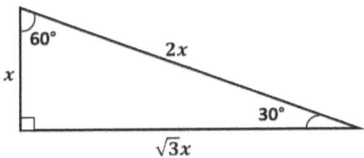

23. GEOMETRY – CIRCLES

Area of the circle = $\boldsymbol{\pi r^2}$

Circumference of the circle = $\boldsymbol{2\pi r}$

Where $\pi = \dfrac{22}{7}$ and r is the radius of the circle.

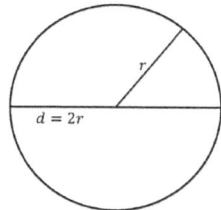

Arc Length = AB = $\dfrac{x}{360} \times \boldsymbol{2\pi r}$

Sector Area = $\dfrac{x}{360} \times \boldsymbol{\pi r^2}$

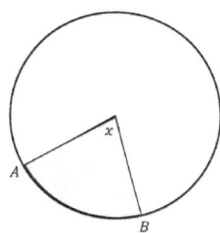

An inscribed triangle with one side passing through the centre is always a right angled triangle. $\Delta\, ABC$ and $\Delta\, ABD$ are both right angles triangles.

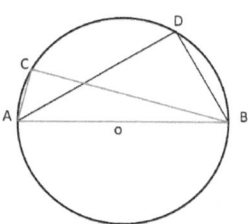

The angle formed at the centre is twice that formed at the circumference, if the end points are

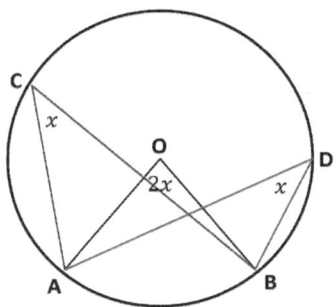
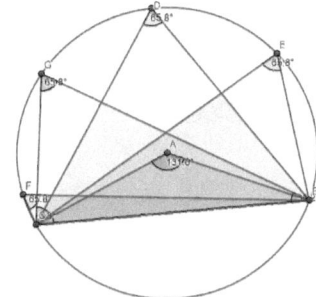

TANGENT

A tangent is a line that touches a circle at just one point. The radius to the point of tangency is always perpendicular to the tangent line.

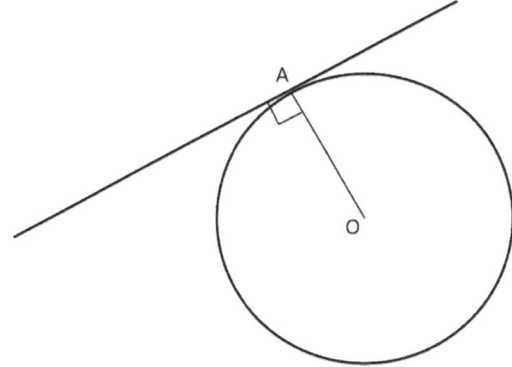

24. GEOMETRY – QUADRILATERALS

A four sided figure is called a quadrilateral.

TRAPEZOID

When two opposite sides are parallel to each other, the shape is called a Trapezoid.

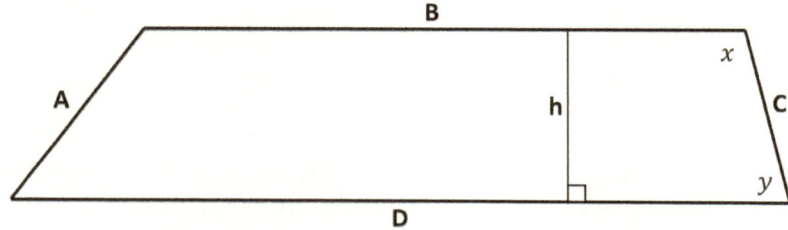

B is parallel to D. The area of the trapezoid is equal to the average of the parallel sides times the perpendicular distance between them.

$$A = \frac{B + D}{2} \times h$$

Adjacent angles of a trapezoid are supplementary.

$$x + y = 180°$$

PARALLELOGRAM

When both opposite pairs of a four sided figure are parallel lines, the shape is called a Parallelogram.

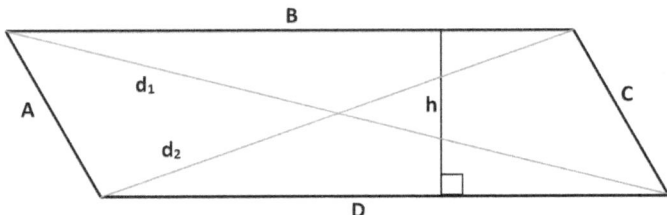

- Opposite sides of a parallelogram are congruent.
- Opposite angles of a parallelogram are congruent.
- Both diagonals bisect each other.
- Consecutive angles are supplementary. $\angle BC + \angle CD = 180°$
- Area of a parallelogram = Side x Height = **Bh**

RECTANGLE

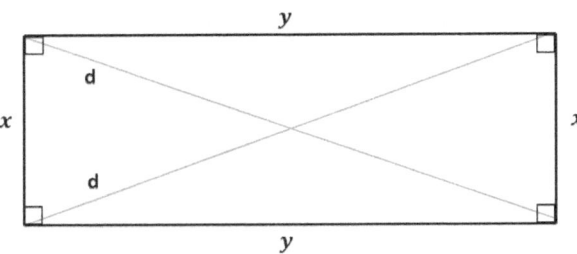

- The opposite sides of a rectangle are parallel and equal to each other.
- All angles are right angles.
- The longer sides are called length and the shorter ones width of the rectangle.
- The diagonals of a rectangle are equal in length.
- The diagonals of a rectangle bisect each other.
- Diagonal = $\sqrt{x^2 + y^2}$
- Area = xy
- Perimeter = $2(x + y)$

SQUARE

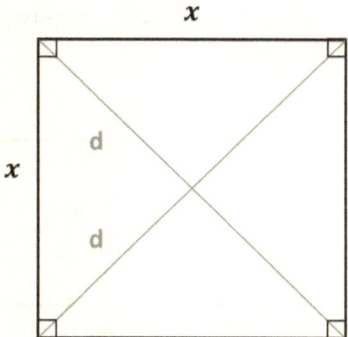

- All sides of a square are equal in length.
- All angles of a square are right angles.
- Diagonals of a square are equal in length.
- Diagonals of a square bisect each other.
- Diagonal = $\sqrt{2}x$
- Area of a square = x^2 or $\dfrac{d^2}{2}$
- Perimeter of a square = $4x$
- A square has the maximum area of a four sided figure with a given perimeter.

25. GEOMETRY – POLYGONS

Quadrilateral – Four sided
Pentagon – Five sided
Hexagon – Six sided
Heptagon – Seven sided
Octagon – Eight sided
Nonagon – Nine sided
Decagon – Ten sided

The sum of the interior angles of a polygon = $180(n - 2)$, where n is the number of sides of the polygon.

The sum of the exterior angles of a polygon = $360°$.

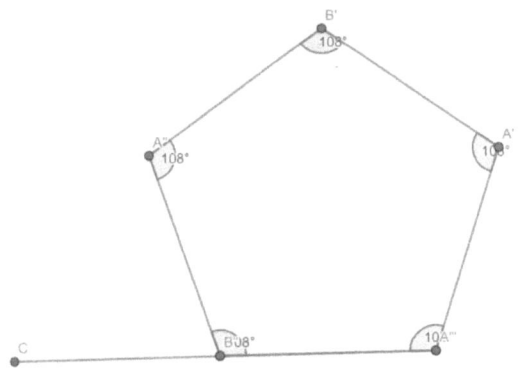

Area of a Regular Pentagon $= \dfrac{1}{4}\sqrt{5+(5+2\sqrt{5})}a^2$, where a is the length of the side of the pentagon.

Diagonal of a Regular Pentagon $= \dfrac{1+\sqrt{5}}{2}a$

Area of a Regular Hexagon $= \dfrac{3\sqrt{3}}{2}a^2$

Area of a Regular Octagon $= 2(1 + \sqrt{2})a^2$

---

26. GEOMETRY – COORDINATE GEOMETRY

Coordinate Geometry is also called Analytic Geometry or Cartesian Geometry.

DISTANCE FORMLUA

Distance between two points A and B is given by

$$AB = \sqrt{(x_2 - x_1)^2 + (y_2 - y_1)^2}$$

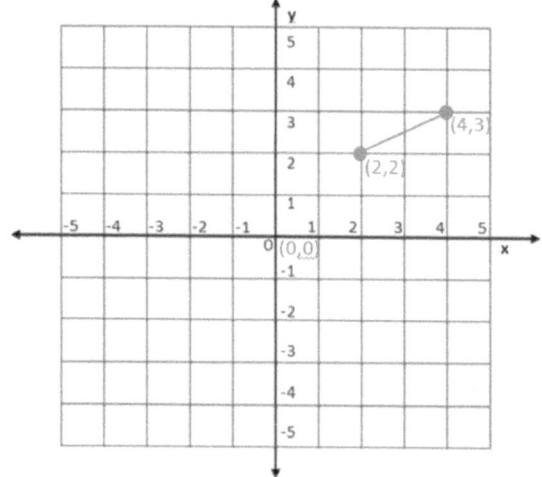

So the distance between A(2,2) and B(4,3) will be

$$AB = \sqrt{(4 - 2)^2 + (3 - 2)^2}$$
$$AB = \sqrt{4+1}$$
$$AB = \sqrt{5}$$

MID-POINT FORMULA

The coordinates of the midpoint, M, between two points A and B can be found by

$$M(x, y) = \left(\frac{x_1 + x_2}{2}, \frac{y_1 + y_2}{2} \right)$$

SLOPE FORMULA

$$Slope = m = \frac{y_2 - y_1}{x_2 - x_1}$$

EQUATION OF THE LINE:

$$y = mx + c$$

Where m is the slope and c is the y-intercept.

Y-INTERCEPT

The point when the line crosses the y-axis.

SLOPES OF PARALLEL LINES:

Slopes of two parallel lines are equal to each other.

$$m_1 = m_2$$

SLOPES OF PERPENDICULAR LINES:

Slope of a line perpendicular to another line is equal to the negative reciprocal of the slope of another line.

$$m_2 = -\frac{1}{m_1}$$

27. GEOMETRY – SOLID GEOMETRY

CUBE:

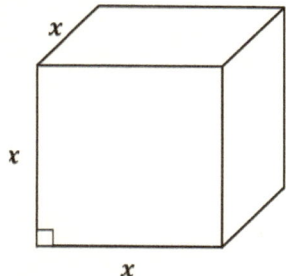

- There are six faces in a cube.
- There are 8 vertices in a cube.
- There are 12 edges in a cube.
- All the faces of a cube are square shaped.
- All angles in a cube are right angles.
- Volume of a cube $= side^3$
- Surface Area of a cube $= 6side^2$
- Longest length is the diagonal $= \sqrt{3}\ side$

CUBOID:

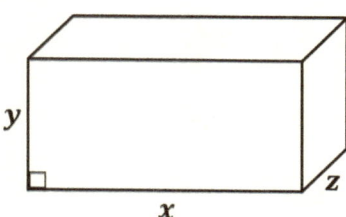

- There are six faces in a cuboid.
- There are 8 vertices in a cuboid.
- There are 12 edges in a cuboid.
- Faces of a cuboid are either rectangular of square.
- All angles in a cuboid are right angles.
- Volume of a cuboid $= \boldsymbol{xyz}$
- Surface Area of a cuboid $= \boldsymbol{2(xy + yz + xz)}$
- Longest length is the diagonal $= \sqrt{x^2 + y^2 + z^2}$

SPHERE:

- Volume of a sphere $= \dfrac{4}{3}\pi r^3$, where r is the radius of the sphere.
- The Surface Area of a sphere $= 4\pi r^2$
- All points on the surface of a sphere are equidistant from the centre.
- A sphere has only one face.
- A sphere is the most efficient container. It holds the maximum volume in a given surface area.

CYLINDER: (Right Circular Cylinder)

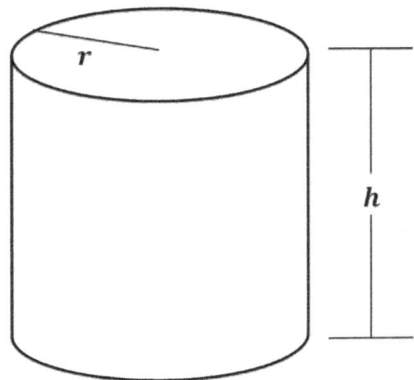

- The two bases of a cylinder are parallel to each other.

- The two bases are congruent.
- Volume of a cylinder = $\pi r^2 h$, where r is the radius of the base and h is the height.
- Surface Area of a sphere = $4\pi r^2$

28. SEQUENCES AND SERIES

Arithmetic Sequence

A set of numbers that are equally spaced, with a fixed difference between two consecutive members, is called an Arithmetic Sequence. Such as

$$10, 13,16, 19, 22,25, ...$$
$$45, 41,37, 33,29, ...$$

1. The Arithmetic Mean, M_n, of n terms of an Arithmetic Sequence is calculated by

$$M_n = \frac{a_1 - a_n}{2}$$

2. Any term of an Arithmetic Sequence can be calculated by using the formula

$$a_n = a + (n - 1)d$$

Where a_n is the nth term, a is the first term, n is the number of terms, and d is the common difference.

3. The sum, S_n, of any n terms of an Arithmetic sequence is given by

$$S_n = \frac{n}{2} \{2a + (n - 1)d\}$$

$$\text{Or } S_n = nM_n$$

4. The sum of first n positive integers is

$$S = \frac{n(n+1)}{2}$$

5. The sum of first n positive odd numbers is equal to n^2.

Geometric Sequence

A sequence of numbers generated after multiplying a common ratio is called a geometric sequence. The numbers grow or diminish exponentially.

1. The nth term of a geometric sequence $= ar^{n-1}$, where a is the first term and r is the common ratio.

2. Sum of the first n terms of a geometric sequence $= a\,\dfrac{1-r^n}{1-r}$

www.ingramcontent.com/pod-product-compliance
Lightning Source LLC
Chambersburg PA
CBHW021513210526
45463CB00002B/995